TRAIN
DRIVER
INTERVIEW
QUESTIONS
& ANSWERS

THE **TESTING** SERIES
expert advice on interview preparation

how2become

Orders: Please contact How2become Ltd,
Suite 2, 50 Churchill Square Business Centre, Kings Hill, Kent ME19 4YU.

Telephone: (44) 0845 643 1299 - Lines are open Monday to Friday 9am until 5pm. Fax: (44) 01732 525965. You can also order via the e mail address info@how2become.co.uk.

ISBN: 978-1907558924

First published 2012

Typeset for How2become Ltd by Molly Hill, Canada.

Printed in Great Britain for How2become Ltd by Bell & Bain Ltd, 303 Burnfield Road, Thornliebank, Glasgow G46 7UQ.

CONTENTS

INTRODUCTION

Dear Sir/Madam,

Welcome to *Train Driver Interview Questions & Answers.* This guide has been designed to help you prepare for and pass the interviews which form part of the Trainee Train Driver selection process. We feel certain that you will find the contents of this invaluable in your pursuit to obtaining one of the most sought after careers available.

The Train Driver selection process is not easy. It is comprehensive, relatively drawn out and highly competitive. In fact, on average there are between 300 and 400 applicants for every vacancy. Coupled with the fact that Train Operating Companies rarely advertise posts, this makes it an even harder job to obtain. However, do not let this put you off as many of the applicants who do apply are grossly under prepared and they normally fail at the first hurdle.

During the selection process you are likely to undergo two different types of interview; the first is a structured/competency based in interview and the second is a manager's interview. Within this guide we will cover both sets of interview.

If you would like any further assistance with the selection process then we offer the following products and training courses via the website:

www.how2become.com:

- How 2 pass the Train Driver interview DVD
- 1 Day intensive Train Driver course
- Train driver testing books and resources.

Finally, you won't achieve much in life without hard work, determination and perseverance. Work hard, stay focused and be what you want!

Good luck and best wishes,

The how2become team

The How2become team

PREFACE BY AUTHOR RICHARD MCMUNN

Before I get into the guide and teach you how to prepare for both sets of train driver interview, it is important that I explain a little bit about my background and why I am qualified to help you succeed.

I joined the Royal Navy soon after leaving school and spent four fabulous years in the Fleet Air Arm branch on board HMS Invincible. It had always been my dream to become a Firefighter and I only ever intended staying the Royal Navy for the minimum amount of time. At the age of 21 I left the Royal Navy and joined Kent Fire and Rescue Service. Over the next 17 years I had an amazing career with a fantastic organisation. During that time I was heavily involved in training and recruitment, often sitting on interview panels and marking application forms for those people who wanted to become Firefighters. I also worked very hard and rose to the rank of Station Manager. I passed numerous assessment centres during my time in the job and I estimate that I was successful at over 95% of interviews I attended.

The reason for my success was not because I am special in anyway, or that I have lots of educational qualifications, because I don't! In the build-up to every job application or promotion I always prepared myself thoroughly. I found a formula that worked and that is what I intend to teach you throughout the duration of this book.

Over the past few years I have taught many people how to pass the selection process for becoming a Trainee Train Driver, both through this guide and also during my one day intensive training course at **www.traindrivercourse.co.uk.**

Each and every one of the students who attends my course is determined to pass, and that is what you will need to do too if you are to be successful. As you are probably aware many people want to become a Train Driver. As a result of this, the competition is fierce. However, the vast majority of people who do apply will submit poor application forms or they will do very little to prepare for the assessment centre and the interviews; as a result, they will fail. If you are at the interview stage then you have done extremely well. The Train Operating Company are interested in seeing you face-to-face and they also want to see what skills and experiences you have which will both match the assessable competencies for becoming a train driver and that will fit into their organisations ethos and values.

The way to pass the train driver interview is to embark on a comprehensive period of intense preparation. I would urge you to use an action plan during your preparation, just like every other area of the selection process. This will allow you to focus your mind on exactly what you need to do in order to pass. If you use an action plan then you are far more likely to achieve your goals.

I use action plans in just about every element of my work. Action plans work simply because they focus your mind on what needs to be done. Once you have created your action plan, stick it in a prominent position such as your fridge door. This will act as a reminder of the work that you need to do in order to prepare properly for selection. Your action plan might look something like this:

My weekly action plan for preparing for Train Driver selection interview

Monday	Tuesday	Wednesday	Thursday	Friday
Research into the TOC I am applying for. Includes reading recruitment literature and visiting websites.	60 minute Interview preparation including preparing my responses to questions.	Carrying out a mock interview with a friend or relative that covers all of the questions within this guide.	Working on my interview technique by sitting in front of a mirror and practicing my responses.	Cross-checking my interview question responses with the assessable qualities for becoming a train driver.

Note: Saturday and Sunday, rest days.

 THE **TESTING** SERIES

The above sample action plan is just a simple example of what you may wish to include. Your action plan will very much depend on your strengths and weaknesses. After reading this guide, decide which areas you need to work on and then add them to your action plan. Areas that you may wish to include in your action plan could be:

- Researching the role of a Train Driver;
- Researching the training that you will undergo as a Trainee Train Driver;
- Researching the Train Operating Company that you are applying for;
- Cross-checking your prepared interview responses with the skills and qualities required to become a train driver;
- Carrying out a mock interview to help you practice your responses to the questions;
- Fitness workouts in order to improve your health, concentration levels and well-being;
- Working on and developing your interview technique.

Finally, it is very important that you believe in your own abilities. It does not matter if you have no qualification; It does not matter if you have no knowledge yet of the role of a Train Driver. What does matter is self-belief, self-discipline and a genuine desire to improve and become successful. If you follow my advice within this guide and use an action plan to keep a track of your progress then your chances of success will increase greatly.

Enjoy reading the guide and then set out on a period of intense preparation!

Best wishes,

Richard McMunn

Richard McMunn
www.TrainDriverCourse.co.uk

CHAPTER 1
ABOUT THE TRAIN DRIVER INTERVIEWS

During this section of the guide I will provide you with an explanation of the different types of interview you may be required to go through. It is important that you check with the Train Operating Company you are applying to which type of interview you will be sitting. Once you have this information you can then extract and use the appropriate sections from this guide during your preparation.

The two types of interview during trainee train driver selection are STRUCTURED/CRITERIA BASED and a MANAGERS INTERVIEW. Let's now take a look at each of them individually in order to give you a better understanding of what is required and how they differ.

THE STRUCTURED/CRITERIA BASED INTERVIEW

The structured interview will come before the manager's interview and is designed to assess whether or not you have the skills, qualities, attributes and experience to become a train driver.

The interview is so designed to provide the assessors with evidence of a candidate's ability, motivation, personal style and attitudes in six key areas which determine how an individual will cope with the unique demands of the driving role and its environment. Prior to the interview you are asked to give relevant examples on a candidate interview form. The interviewer then uses this form as a basis for an in-depth exploration of your behaviour. The interviewer/assessor will then record the evidence you provide on a structured interview record form.

The Train Driver structured interview is based on thorough job analysis of the UK train driver role. The interview process has been used successfully within the UK for over 12 years and reveals invaluable evidence about a candidate's approach to the role.

During your preparation I recommend that you think about your experiences so far in life in the following areas:

- Your ability to work well within a team;
- Your ability to work effectively under pressure;
- Your ability to learn and retain large amounts of job-related information;
- Your ability to provide a high level of customer service;
- Your ability to cope with urgency under pressure;
- Your ability to follow policies and methods;
- Your ability to work by yourself, unsupervised;
- Your ability to communicate effectively;
- Your ability to remain attentive for long periods of time.

There are of course other key areas to the train driver's role; however, the above key areas are an ideal starting point for your preparation.

Here are a few examples of structured interview questions:

SAMPLE STRUCTURED/CRITERIA BASED INTERVIEW QUESTIONS

Q. Provide an example of when you have worked as part of a team. Within your response please explain what you did and how you contributed towards the team.

Q. Provide an example of when you have had to communicate an important message to a group of people.

Q. Provide details of where you have carried out a safety-critical task.

Q. Provide an example of when you have delivered high-quality customer service.

Q. Provide details of a situation where you had to be attentive for long periods of time whilst working on a difficult task.

Q. Provide details of when you have learnt a large amount of job-related information.

Q. Provide details of when you have had to carry out a task in a highly pressurised environment.

Q. Provide an example of when you have carried out a difficult task unsupervised.

Q. Provide an example of when you have followed policies and set methods to complete a specific task.

Q. Provide details of when you have been in a situation of urgency and under pressure. What did you do and why?

The above sample structured interview questions all require you to provide a specific example of where you carried out a task or took a particular course of action. You will notice that the sample questions are all within the realms of the role of a train driver; therefore, it is perfectly acceptable for the interviewing panel to want to see EVIDENCE of where you meet each particular area being assessed. You will notice that I have capitalised the word EVIDENCE; this is because it is vitally important that you provide specific details of where you have already carried out the task or action in question in a previous or current role. Anybody can say at interview what they would do in a given situation, but actually providing details and EVIDENCE of what they have done in a similar situation is a far harder task.

As you can imagine, many people worry that they do not have the life experiences to answer these types of questions. I get lots of people ask me during my 1-day train driver course how do they obtain the experience to answer these questions. My simple answer is that they need to go out there and get the experience necessary. Of course, being a former firefighter I personally have lots of experiences in all of the above areas; however, I sympathise with those of you who find it difficult to think of situations that you have been in where you can provide EVIDENCE of the assessable area. The simple fact is this: you are applying for a position that carries a huge amount of responsibility and as such, it is vital that you have the knowledge, skills and experiences to do the job safely and competently. If you don't currently have the experiences to answer the questions then you need to find ways of gaining the required level of experience. For example, there is nothing to stop you from embarking on a Health and Safety course in order to gain the skills and experiences required to answer the following two questions:

Q. Provide an example of when you have followed rules and procedures to complete a specific task.

Q. Provide details of where you have carried out a safety-critical task.

There is also nothing to stop you from asking your boss at work if you can deliver a presentation to your co-workers on an interesting subject that relates to your work. By doing this you will be able to answer the following question:

Q. Provide an example of when you have had to communicate an important message to a group of people.

If you are struggling to think of a situation where you have learnt a large amount of work-related information then why not embark on an evening class or learn a musical instrument?

The point I am making here is that there are ways in which you can gain experiences in order to answer the questions; you just need to think outside of the box a little!

The important factor to remember of that at no point should you be dishonest during the interview. People have asked me during my 1-day train driver course if it is wise for them to 'make up' fictitious scenarios so that they can answer the questions; my response is a resounding NO! There is absolutely no way that I want a train driver driving a train that I am a passenger on if he or she has lied at the interview; would you?

SAFETY, SAFETY, SAFETY!

The role of a train driver is one that is called in the industry 'safety-critical'. The criteria for what makes a job safety-critical will depend upon the regulatory environment and/or the safety standards applied to the role.

The general consensus is that a job is safety-critical if it may lead to harm being caused (or not being prevented) to users, bystanders, other stakeholders, or the environment, although the range of causes and harm considered can vary. Basically, in a nutshell, the role of train driver is a highly responsible one which requires an ability to carry out the role safely, competently and professionally. If the job is not carried out in line with rules and safety procedures then harm may be caused.

During your preparation for both sets of interviews I recommend that you have the word safety at the forefront of your mind at all times. As a firefighter I was required to carry out safety-critical tasks every day of my career; therefore, I am used to operating in a safe manner whilst going about my day-to-day business. Having personally attended and passed a number of Health and Safety courses during my time I can strongly recommend that you consider attending one. By attending and passing a Health and Safety course you will improve your knowledge and experience of this very important subject. You will also find it far easier to answer some of the questions during the interview, too!

THE MANAGER'S INTERVIEW

The manager's interview usually comes after the structured/competency based interview. The train operating company hold this type of interview so that they can see what you are really like as a person. The manager's interview is more like a traditional interview and will most probably cover the following areas:

- The reasons why you want to become a train driver;
- What you already know about the role of a train driver, the working conditions and the training you will undergo;
- What you know about the Train Operating Company (TOC) you are applying to join;
- Why you want to join that particular TOC;
- The qualities you can bring to the role;
- Your motivations and ambitions for becoming a train driver.
- What you would do in given situations.
- How you would get to work at 3am in the morning.
- How flexible you are.

There will of course be other areas that will be covered during the manager's interview; however, if you concentrate on the above key areas during your preparation you will be very well prepared. In addition to being assessed on the above areas you will also be assessed on your style of communication and your interview technique. It is more than likely that there will be two people sitting on the manager's interview. This will be the Human resources manager for the train operating company and also the station manager for whom you will be working under.

PROBING QUESTIONS

Probing questions will be utilised by the interview panel during the structured interview in particular, to explore your responses in more details. Once you have provided your response to the interview questions posed you may get asked further questions such as:

- Why did you act in that particular manner?

- What made you decide to take that course of action?

- Would you do anything differently if the same situation arose again?

- What did you take into consideration?

- How did the situation make you feel?

When preparing for the interview questions in a later chapter of this guide think carefully about the types of probing questions you could get asked and have a response prepared for them.

I also strongly believe it is good practice to review reflect on your performance after every work-related task that you carry out and I would encourage you to use this phrase when responding to some of the interview questions. For example, you may decide to say something like:

"After I had completed the task I sat down and carried out a review of my work and how I had reached my goal. I identified that there were a number of ways that I could improve for next time. I feel it is good practice to review and reflect on ones performance after carrying out a task."

Before I provide you with the all-important sample interview questions and tips on how to answer them, in the next chapter I will provide you with some essential information on how to improve your technique. Many of you might be tempted to skip out this section and move on to the

interview questions; please don't. Read the information as I promise it will help you to gain higher marks during both sets of interviews.

CHAPTER 2
HOW TO IMPROVE YOUR INTERVIEW TECHNIQUE

The Train Driver Interview does not have to be a daunting process, providing that is, you prepare effectively. Yes, any interview can be a nerve-wracking experience, but if you prepare in the right areas this will give the confidence you need to pass with flying colours.

Within this section of the guide I will explain some of the more important aspects of interview technique you should focus on. It may feel uncomfortable but I would strongly advise that you carry out a mock interview with a friend or relative before the actual interview. Get them to ask you the questions that are contained within this guide. This will give you the opportunity to work on your responses as well as your technique. If you have the time, I would also practice sitting in front of a mirror and watching yourself answer the questions. This will give you a good idea of how you come across during an interview.

When practising your interview technique you should:

- Sit up right in the chair;
- Speak clearly;

- Dress appropriately for the occasion;

- Never slouch or look dis-interested;

- Answer the questions in a concise and logical manner;

- Smile and be enthusiastic.

Let us now take a look at how to can prepare for the interview.

HOW TO PREPARE EFFECTIVELY

During your preparation for the interview I would recommend that you concentrate on the following three key areas:

- Interview technique;

- Research;

- Responding to the interview questions.

Each of the above areas is equally important. I will now go into each one of them in detail:

INTERVIEW TECHNIQUE

Interview technique covers a number of different areas. The majority of candidates will pay it little, if any attention at all. Interview technique basically involves the following key areas:

> **Creating the right impression**. When you walk into the interview room you should stand up tall, smile and be polite and courteous to the panel. Do not sit down in the interview chair until invited to do so.

> **Being presentable.** During my time as an interviewer for a number of different jobs I have been amazed at the number of people who turn up inappropriately dressed. I have seen people turn up for interviews in jeans, t shirts and trainers! I strongly advise that you take the time to look smart and presentable. Remember you are applying to join an organisation that requires you to wear a uniform. If you dress smart and formal for the interview then you are far more likely to wear your uniform with pride. As a train driver you are a role model for the company.

> **Sitting in the chair.** The interview could last for up to an hour, depending on the length of your responses to the questions. This is a long time to concentrate for. Whilst in the interview chair sit up right at all times and never slouch.

> **Motivation.** Throughout the duration of the interview demonstrate a high level of motivation and enthusiasm. You do not want to come across as desperate, but conversely you must come across as highly motivated and determined to be successful. Always smile and be respectful of the interview panel.

> **Communication.** When communicating with the interview panel look them in the eye but never stare at them in an intimidating manner. This shows a level of confidence. You should also communicate in a clear and concise manner where possible.

> **Asking questions.** At the end of the interview you will be given the opportunity to ask questions. This is where some candidates let themselves down with silly or inappropriate questions that relate to leave or sick pay. It is quite acceptable to ask a couple of questions, however, keep them simple and relevant. Examples of good questions to ask are:

Q. If I am successful, how long would it be before I start my train driver training?

Q. I have been looking into your company and I have been impressed with the 'meet the manager's' scheme that you operate for your customers. Has this been successful?

Q. Whilst I am waiting to find out if I am successful, are there any documents I could read to learn more about the company and the role of a train driver?

> **A final parting statement.** Once the interview has finished and you have asked your questions, you may wish to finish off with a final statement. Your final statement should say something about your desire and passion for becoming a Trainee Train Driver. The following is a good example of a final statement:

"I would just like to say thank you for giving me the opportunity to be interviewed for the post today. Over the last few months I have been working hard to learn about the role and also about your company. If I am successful then I promise you that I will work hard to pass the tests and exams and I will be a loyal and professional employee of your team. Thank you."

RESEARCH

As you can imagine in the build-up to the interview you will need to carry out plenty of research. Research that is, in relation to the role of a Trainee Train Driver and also the Train Operating Company that you are applying to join. Here is a list of the more important areas I recommend that you study:

- The job description and person specification for the job that you are applying for.

- Your application form and the responses that you provided to all of the questions.

- The website of the Train operating Company you are applying to join. What is their customer service charter? Do they have a mission statement? What are their core values? What services do they provide? What is their geographical area? How many people work for them? Who is the person in charge? What stations do they operate out of? What trains do they operate? Do they operate any schemes in order to improve customer service? What are the future plans of the TOC?

- Try to visit a train station that the TOC operates out of. Speak to some of the staff at the station and ask them questions about the role they perform. Try to find out as much as possible about the TOC you are applying for. If you get the opportunity, speak to a qualified Train Driver who works for the TOC. You may also decide to telephone the TOCs Human Resources department and ask if you can go along to find out a little bit more about their organisation and what they expect from their employees.

RESPONDING TO THE INTERVIEW QUESTIONS

If I was preparing for the Trainee Train Driver interview right now I would take each area of the role individually and prepare a detailed response setting out where I meet the requirements of it.

Your response to each question that relates to the role of a Trainee Train Driver must be 'specific' in nature. This means that you need to provide an example of where you have already demonstrated the skills that are required under the job description or person specification in a previous role or situation. This is particularly relevant to questions that are asked during the structured/competency based interview.

Do not fall into the trap of providing a 'generic' response that details what you 'would do' if the situation arose. Try to structure your responses in a logical and concise manner. The way to achieve this is to use the 'STAR' method of interview question response construction:

Situation

Start off your response to the interview question by explaining what the 'situation' was and who was involved.

Task

Once you have detailed the situation, explain what the 'task' was, or what needed to be done.

Action

Now explain what 'action' you took, and what action others took. Also explain why you took this particular course of action.

Result

Finally explain what the outcome or result was following your actions and those of others. Try to demonstrate in your response that the result was positive because of the action you took.

You should also consider what you would do differently if the same situation arose again. It is good to be reflective at the end of your responses. This demonstrates a level of maturity and it will also show the panel that you are willing to learn from every experience.

On the following pages I have provided you with a number of sample interview questions and responses to assist you in your preparation. Please remember that the responses provided are not to be copied under any circumstances. Use them as a basis for your preparation taking examples from your own individual experiences and knowledge.

You will also notice that I have supplied a blank space after some of the questions. This is to help you create your own response for that particular question based on your own skills, qualifications and experiences. I strongly suggest you take the time to write down your own answers as this will help you to prepare better than the other candidates.

I have also indicated which interview I believe the sample question will be asked under. I cannot guarantee that these questions will come up for each interview but these are the ones that I would personally prepare for.

CHAPTER 3
SAMPLE INTERVIEW QUESTIONS

Question 1

Why do you want to become a Train Driver? (Manager's interview)

This question is inevitable, so it is important that you ensure you have a suitable answer prepared. Many people will respond with a standard answer such as *"It's something that I've always wanted do since I was young"*. Whilst this is OK you need to back it up with genuine reasons that relate to the TOC you are applying for and other important reasons such as working in a customer-focused environment and a desire to learn new skills.

This type of question may be posed in a number of different formats such as the following:

Q. Why do you want to become a Train Driver with our Company?

Q. What has attracted you to the role of Train Driver?

Now take a look at the following sample response which will help you to prepare for this type of question. Once you have read it, use the template on the next page to create your own response based upon your own experiences and knowledge.

Sample Response

Why do you want to become a Train Driver?

"I have wanted to become a Train Driver for many years now and have been preparing for the role for a long time. I have been very careful about which TOC to apply for and I have been impressed with the way your company operates. It sets itself high standards in terms of customer service and the safety standards that are expected of its employees. Apart from the fact that driving trains is quite an exciting job, I also very much enjoy new and different challenges. I understand that as a Train Driver there are a lot of new skills to learn, especially during the early years. The type of person I am means that I would work hard to ensure that I passed every exam first time. I also enjoy working in a customer-focused environment where a high level of service is essential. As a Train Driver you are responsible for the customer's safety and I would enjoy the high level of responsibility that comes with the position."

Template for Question 1

WHY DO YOU WANT TO BECOME A TRAIN DRIVER?

HAVING WORKED IN THE AIRLINE INDUSTRY
FOR MANY YEARS I HAVE BEEN LOOKING FOR
A NEW CHALLENGE AND CHANGE IN CAREER.
I REALISED THAT MANY OF THE SKILLS I HAVE
LEARNT ARE SIMILAR TO THAT OF A TRAIN DRIVER.
AS CABIN CREW I WORK IN A CUSTOMER FOCUSED
AND SAFETY CRITICAL ROLE. WHICH HAS ALLOWED
ME TO BECOME A GREAT COMMUNICATOR, WORK
CALMLY IN HIGHLY PRESSURRED AND CRITICAL
SITUATIONS, WHILST FOLLOWING RULES AND
PROCEDURES ENABLING ME TO OPERATE SAFELY
AT ALL TIMES. I UNDERSTAND ABELLIO SCOTRAIL
WILL BE INVESTING A LOT OF TIME AND
MONEY INTO MY TRAINING IF I WAS TO BE
SUCCESSFUL AT THIS INTERVIEN. I AM RELIABLE
AND AN EXPERIENCED TEAM PLAYER AND
WILL WORK HARD.

Question 2

Why do you want to work for our company? (Manager's interview)

Once again this is a question that is likely to come up during your interview. In order to prepare for this question you need to carry out some research about the TOC you are applying for. The best place to get this information is via their website. See the Useful Contacts section for a list of current TOCs.

When responding to this type of question, try to focus on the positive aspects of the company's work. Do they run any customer-focused initiatives or have they won any awards for quality of work or service? It is always good to make reference to the positive aspects of their work, but do not make mention of any current or previous bad press. On the following page I have provided a sample response to this question to help you prepare. I have used Southern Rail as an example when constructing the response. Once you have read it, take the time to construct your own answer using the template provided.

Sample Response

Why do you want to work for our company?

"I have been looking at a number of TOCs and I have been especially impressed with Southern Rail. The 'Meet the Managers' programme gives passengers the chance to meet Senior Managers and Directors and talk with them about the service. This demonstrates a high level of customer focus and care and I want to work for such a company as I believe I can bring the same high standards to the team.

I also understand that, over the next two years, Southern Rail aims to create a company that not only looks and feels different, but provides passengers with a better travelling experience. I believe that, whilst working with Southern Rail, I would have excellent career opportunities and therefore be very happy in my role as a Train Driver."

Template for Question 2

WHY DO YOU WANT TO WORK FOR OUR COMPANY?

_____

_____

_____

_____

_____

_____

_____

_____

_____

_____

_____

_____

_____

_____

_____

_____

_____

_____

_____

_____

_____

_____

_____

_____

Question 3

What can you tell us about the role of a Train Driver? (Manager's interview)

You must be well prepared for this question prior to your interview. If you don't know what the role involves, then you shouldn't be applying for the post. When responding to this question, make sure you make reference to the job/person specification for the role. The job specification is a 'blue-print' for the role that you will be required to perform whilst working as a Train Driver. Therefore, it is essential that you know it. An example of a Train Driver's duties/person specification is detailed below:

Person specification

We need people who will share our passion, enthusiasm and commitment to deliver the best train service for our passengers. You will have outstanding communication skills and be strong on customer service. You must be able to learn and follow rigorous railway and safety procedures. Aged between 21-49, you will need to be decisive and have excellent concentration skills.

Job description

Drive trains taking particular account of all permissible, temporary and emergency speed restrictions. Trains must be driven with regard to punctuality and customer comfort. Obey all fixed signals and hand signals. Communicate effectively with signallers and hand signallers regarding the transmission of verbal messages. The use of phonetic alphabet is mandatory. Ensure that customers are advised, either directly or through others, regarding train running matters. Ensure that your traction and route cards are kept up to date.

Now take a look at the sample response on the following page before constructing your own response using the template provided.

Sample Response

What can you tell us about the role of a Train Driver?

"I understand that the role involves a high level of responsibility, concentration and lone working. To begin with, Train Drivers are responsible for ensuring that they drive trains taking particular account of all permissible, temporary and emergency speed restrictions.

The trains must be driven on time and drivers must ensure that they obey all fixed signals and hand signals as the safety of the trains and passengers is paramount. Part of the role also involves communicating with signallers and hand signallers regarding the transmission of verbal messages. Other elements of communication involve ensuring that customers are advised, either directly or through others, regarding train running matters.

Finally, Train Drivers must ensure that their traction and route cards are kept up to date. Safety is essential to the role of a competent Train Driver and keeping up to date with procedures and regulations is very important."

Template for Question 3

WHAT CAN YOU TELL US ABOUT THE ROLE OF A TRAIN DRIVER?

Question 4

What skills do you possess that you think would be an asset to our team? (Manager's interview)

When responding to questions of this nature, try to match your skills with the skills that are required of a Train Driver. On some TOC websites, you will be able to see the type of person they are looking to employ, usually in the recruitment section.

An example of this would be*: 'We are looking for friendly, supportive people who share our professional, customer-focused approach. You must be a good team player with a flexible attitude and a willingness to learn.'* Just by looking at the TOC's website, you should be able to obtain some clues as to the type of person they are seeking to employ. Try to think of the skills that are required to perform the role you are applying for and include them in your response.

The following is a sample response to the question. Once you have read it, take the time to construct your own response using the template provided.

Sample Response

What skills do you possess that you think would be an asset to our team?

"I am a very conscientious person who takes the time to learn and develop new skills correctly. I have vast experience working in a customer-focused environment and fully understand that customer satisfaction is important. Without the customer there would be no company, so it is important that every member of the team works towards providing a high level of service.

I believe I have the skills, knowledge and experience to do this. I am a very good team player and can always be relied upon to carry out my role to the highest of standards. I am a flexible person and understand that there is a need to be available at short notice to cover duties if required. In addition to these skills and attributes, I am a very good communicator. I have experience of having to communicate to customers in my previous role and believe that this would be an asset in the role of a Train Driver. I am highly safety conscious and have a health and safety qualification to my name. Therefore, I can be relied upon to perform all procedures relevant to the codes of conduct and will not put myself or others in any danger whatsoever. Finally, I am very good at learning new skills which means that I will work hard to pass all of my exams if I am successful in becoming a trainee Train Driver."

Template for Question 4

WHAT SKILLS DO YOU POSSESS THAT YOU THINK WOULD BE AN ASSET TO OUR TEAM?

Question 5

Can you tell us about a situation when you have had to work under pressure? (Structured interview)

The role of a Train Driver will sometimes involve a requirement to work under pressure. Therefore, the recruitment staff want to know that you have the ability to perform in such an environment. If you have experience of working under pressure then you are far more likely to succeed as a Train Driver. When responding to a question of this nature, try to provide an actual example of where you have achieved a task whilst being under pressure. Questions of this nature are sometimes included in the application form, so try to use a different example for the interview.

I have provided you with a sample response to this question. Once you have read it, take the time to construct your own response based on your own individual experiences and knowledge using the template provided.

Sample Response

Can you tell us about a situation when you have had to work under pressure?

"Yes, I can. In my current job as car mechanic for a well- known company, I was presented with a difficult and pressurised situation. A member of the team had made a mistake and had fitted a number of wrong components to a car. The car in question was due to be picked up at 2pm and the customer had stated how important it was that his car was ready on time because he had an important meeting to attend.

We only had two hours in which to resolve the issue and I volunteered to be the one who would carry out the work on the car. The problem was that we had three other customers in the workshop waiting for their cars too, so I was the only person who could be spared at that particular time. I worked solidly for the next two hours making sure that I meticulously carried out each task in line with our operating procedures. Even though I didn't finish the car until 2.10pm, I managed to achieve a very difficult task under pressurised conditions whilst keeping strictly to procedures and regulations."

Template for Question 5

CAN YOU TELL US ABOUT A SITUATION WHEN YOU HAD TO WORK UNDER PRESSURE?

_____

_____

_____

_____

_____

_____

_____

_____

_____

_____

_____

_____

_____

_____

_____

_____

_____

_____

_____

_____

_____

_____

Question 6

Can you tell me about a time when you have worked as part of a team to achieve a goal? (Structured interview)

Having the ability to work as part of a team is very important to the role of a Train Driver. Train Operating Companies employ many people in different roles from Conductors to platform staff and from ticket office staff to caterers. In fact it is not uncommon for thousands of people to work for one particular TOC. Therefore, it is essential that every member of the team works together in order to achieve the ultimate goal of providing a high quality rail service.

The recruitment staff will want to be certain that you can work effectively as part of a team, which is why you may be asked questions that relate to your team working experience.

There now follows a sample response to this question. Once you have read it, take time to construct your own response using the template provided.

Sample Response

Can you tell me about a time when you have worked as part of a team to achieve a goal?

"Yes, I can. I like to keep fit and healthy and as part of this aim I play football for a local Sunday team. We had worked very hard to get to the cup final and we were faced with playing a very good opposition team who had recently won the league title. After only ten minutes of play, one of our players was sent off and we conceded a penalty as a result. Being one goal down and 80 minutes left to play we were faced with a mountain to climb. However, we all remembered our training and worked very hard in order to prevent any more goals being scored. Due to playing with ten players, I had to switch positions and play as a defender, something that I am not used to. The team worked brilliantly to hold off any further opposing goals and after 60 minutes we managed to get an equaliser. The game went to penalties in the end and we managed to win the cup. I believe I am an excellent team player and can always be relied upon to work as an effective team member at all times. I understand that being an effective team member is very important if the Train Operating Company is to provide a high level of service to the passenger. However, above all of this, effective teamwork is essential in order to maintain the high safety standards that are set."

Template for Question 6

CAN YOU TELL ME ABOUT A TIME WHEN YOU HAVE WORKD AS PART OF A TEAM TO ACHIEVE A GOAL?

THE **TESTING** SERIES

Question 7

Can you provide us with an example of a project you have had to complete and the obstacles you had to overcome? (Structured interview)

Having the ability to complete tasks and projects successfully demonstrates that you have the ability to complete your trainee Train Driving course. Many people give up on things in life and fail to achieve their goals. The recruitment staff need to be convinced that you are going to complete all training successfully and, if you can provide evidence of where you have already done this, then this will go in your favour.

When responding to this type of question, try to think of a difficult, drawn out task that you achieved despite a number of obstacles that were in your way. You may choose to use examples from your work life or even from some recent academic work that you have carried out. Take a look at the following sample question before using the template provided to construct your own response based on your own experiences.

Sample Response

Can you provide us with an example of a project you have had to complete and the obstacles you had to overcome?

"Yes I can. I recently successfully completed a NEBOSH course (National Examination Board in Occupational Safety and Health) via distance learning. The course took two years to complete in total and I had to carry out all studying in my own time whilst holding down my current job.

The biggest obstacle I had to overcome was finding the time to complete the work to the high standard that I wanted to achieve. I decided to manage my time effectively and I allocated two hours every evening of the working week in which to complete the work required. I found the time management difficult but I stuck with it and I was determined to complete the course. In the end I achieved very good results and I very much enjoyed the experience and challenge. I have a determined nature and I have the ability to concentrate for long periods of time when required. I can be relied upon to finish projects to a high standard."

Template for Question 7

CAN YOU PROVIDE US WITH AN EXAMPLE OF A PROJECT YOU HAD TO COMPLETE AND THE OBSTACLE YOU HAD TO OVERCOME?

Question 8

Can you provide us with an example of a safety-related task that you have had to perform? (Structured interview)

Safety is an extremely important part of the Train Driver's role, and the recruitment staff need to know that you are capable of working safely at all times. The term 'safety' should be an integral part of your responses during the interview. Making reference to the fact that you are aware of the importance of safety at every opportunity is a positive thing. When responding to safety-related questions try to include examples where you have had to work to, or follow, safety guidelines or procedures. If you have a safety qualification then it is definitely worthwhile mentioning this during your interview. Any relevant safety experience or related role should also be discussed.

Now take a look at the following sample response before using the template provided to construct your own response.

Sample Response

Can you provide us with an example of a safety-related task that you have had to perform?

"I currently work as a gas fitter and I am often required to perform safety-related tasks. An example of one of these tasks would involve the installation of gas-fired boilers. When fitting a gas boiler I have to ensure that I carry out a number of safety checks during the installation stage which ensures my work is safe and to a high standard. I have been trained, and I am qualified, to carry out my work in accordance with strict safety guidelines. I also have a number of safety certificates to demonstrate my competence.

I am fully aware that if I do not carry out my job in accordance with safety guidelines there is the possibility that somebody may become injured or even killed."

Template for Question 8

CAN YOU PROVIDE US WITH AN EXAMPLE OF A SAFETY RELATED TASK THAT YOU HAVE HAD TO PERFORM?

Question 9

Can you give an example of when you have had to work on your own for long periods? (Structured interview)

Lone working is an unfortunate part of the Train Driver's job. You will spend many hours on your own and this can be a problem for many people. You need to think carefully about this downside to the job. Can you cope with it? Do you have any experience of working on your own? If you do not then you will have to convince the panel that you can cope with it.

Sample Response

Can you give an example of when you have had to work on your own for long periods?

"Yes, I have worked on my own unsupervised on numerous occasions. Most recently I was asked by the foreman on a building site to install new gas boilers in four properties within a tight deadline. Whilst I understood it was important to carry out the task quickly, there was no way I was going to compromise on safety. I started off by creating a mini action plan in my head which detailed how I would achieve the task. I set about installing the first boiler conscientiously and carefully whilst referring to the safety manual when required. I made sure that there was sufficient ventilation in the houses as required under health and safety law. During the week that I was required to complete the task I had previously arranged to go to a birthday party with my wife, but I decided to cancel our attendance at the event as I needed to get a good night's sleep after each hard day's work. I knew that if I was to maintain the concentration levels required to work safely and achieve the task then I would need to be in tip-top condition and getting sufficient rest in the evenings was an important part of this. By the end of the fourth day I had successfully completed the task that was set by the foreman and the proceeding safety checks carried out by the inspector on the boilers proved that I had done a very good job."

Question 10

What is your sickness record like and what do you think is an acceptable level of sickness? (Manager's interview)

Most employers detest sickness and they especially detest sickness that is not genuine. For every day that an employee is off sick will cost the TOC dearly. Therefore, they want to employ people who have a good sickness record. Obviously you cannot lie when responding to this question as the TOC will carry out checks. The latter part of the question is simple to answer. Basically no amount of sickness is acceptable but sometimes genuine sickness cannot be helped. Remember to tell them that you do not take time off sick unless absolutely necessary and you can be relied upon to come to work.

Question 11

Have you ever worked during the night and how do you feel about working shifts? (Manager's interview)

Train Driving involves irregular shifts and the Train Operating Company want to know that you can handle them. Speak to any person who works shifts and they will tell you that after a number of years they can start to take their toll. Remember to tell the panel that you are looking forward to working shifts and in particular night duties. If you can provide examples of where you have worked irregular shift patterns then remember to tell them.

Question 12

Would you get bored of driving the same route day in, day out? (Manager's interview)

Of course the only answer here is no! Yes, we would all probably get bored of the same journey every day, but don't tell them this.

Question 13

How many people work for this TOC? (Manager's interview)

Questions that relate to facts and figures about the TOC might come up. They want to know that you are serious about joining them and that you are not just there to become a Train Driver. Make sure you study their website and find out as much about them as possible.

Question 14

How many stations does the company service? (Manager's interview)

Once again, this is a question that relates to your knowledge of the TOC. This kind of information can usually be found by visiting their website. Please see our Useful Contacts section for more details.

Question 15

What are the mission and aims of this company? (Manager's interview)

Many organisations including Train Operating Companies set themselves aims and objectives. They usually relate to the high level of customer service that they promise to deliver. When you apply to become a Train Driver you should not only prepare for each stage of the selection process but you should also learn as much as possible about the company you are applying to join. Learning this kind of information is important and it will demonstrate your seriousness about joining their particular company. Always remember this rule, working for the TOC comes first, becoming a Train Driver comes second! Visit the website of the TOC in order to view their mission, aims, objectives or customer charter.

Question 16

Can you provide us with an example of when you have had to work in an emergency situation? (Structured interview)

This question is also likely to be asked during the application form stage of the process. Being able to remain calm under pressure is very important and will form an integral part of your training. Maybe you have had to deal with an emergency at work or even in the home? Whatever example you decide to use, make sure you tell them that you stayed calm and focused on the task in hand. Make reference to the importance of safety during your response too.

Sample Response

Can you provide us with an example of when you have had to work in an emergency situation? (Structured interview)

"Whilst driving home from work one day I came across a road collision. I checked to see that it was safe to pull over before parking my car just past the accident and with my hazard warning lights on. I then got out of the car from the passenger side as the traffic was still passing quite fast.

As I was approaching the accident I called 999 and requested the Police and Ambulance service as I could see there was a person in the car with blood coming from her head. I gave as much detail to the call operator in relation to the location of the accident and who/what was involved. I remained totally calm and in control throughout. Once I knew that the emergency services were on their way I tried to communicate with the injured lady. I carefully opened the door of the car and stabilised her head to prevent any further neck or spinal injuries. I kept speaking to her whilst we waited for the ambulance crews and the police to arrive. Once the crews arrived I handed the casualty over to them and explained what I had done. The police then took my details and thanked me for my calm approach."

Question 17

Do you think it's important for staff to wear a uniform? (Manager's interview)

The answer to this question should be yes. The reason for this is that a uniform gives customers and passenger's confidence in the service they are receiving. It is also important during an emergency situation so that customers know who to turn to for help and guidance. Uniforms are positive for the image of Train Operating Companies which is why they use them. Be positive about uniforms and tell them that you are looking forward to wearing one and taking pride in your appearance.

Question 18

If you were a train driver and you came across an obstacle on the track, what would you do and why? (Manager's interview)

This type of question is rare but it has been asked during the selection interview. Basically, you need to ensure that you tell the panel that you would follow the procedures you learnt during your training. You should tell them that the safety of your passengers and the train would be paramount. You would raise the alarm immediately by contacting the control centre so that other trains could also be informed of the danger and they could then take appropriate action as necessary. During your responses you should never compromise safety!

Question 19

Tell me about a time when you had to follow clear instructions or rules in order to complete a task? (Structured interview)

As a train driver you will be required to follow instructions on a daily basis. Here's how to structure your response to this question:

• What was the work you were doing?

• What were the rules or instructions that you had to follow?

- What did you do to complete the work as directed?

- What was the result?

- How did you feel about completing the task in this way?

Train operating companies strive for excellence in everything they do. Therefore, it is crucial that you provide a response that demonstrates you too can deliver excellence and maintain high standards. Try to think of a situation, either at work or otherwise, where you have achieved this. Make your response specific in nature. If you have had to follow specific instructions, rules or procedures then this is a good thing to tell the panel.

Question 20

Tell me about a time when you sought to improve the way that you or others do things? (Structured interview)

Part of the train driver's role includes continuous improvement and being able to adapt to change in both rules and company policies. You will be required to continually learn and improve in your role; therefore, the interview panel will want to see clear evidence of where you have already done this in a previous role.

Here are a few pointers on how to structure your answer:

- What was the improvement that you made?

- What prompted this change?

- What did you personally do to ensure that the change was successful?

- What was the result?

Question 21

Tell me about a time when you have taken it upon yourself to learn a new skill or develop an existing one?

Trainee train drivers are required to learn new skills every day during their training. They will attend on-going training courses and they will also read up on new procedures and policies that relate to the rail industry.

In order to maintain a high level of professionalism train drivers must be committed to continuous development. Try to think of an occasion when you have learnt a new skill, or where you have taken it upon yourself to develop your knowledge or experience in a particular subject. Follow the above structure format to create a strong response.

How to structure your response:

- What skill did you learn or develop?
- What prompted this development?
- When did this learning or development occur or take place?
- How did you go about learning or developing this skill?
- What was the result?
- How has this skill helped you since then?

Question 22

What do you understand about the term Health and Safety and who is responsible for it? (Manager's interview)

Health and Safety plays a very important part in the train driver's working day.

As a train driver you will be acutely aware of Health and Safety and how it affects you and your colleagues. Health and Safety is the responsibility of everybody at work. You are responsible for the safety of yourself, the safety of your passengers and for the safety of everybody else.

Health and Safety within the rail industry is governed by the Health and Safety at Work Act 1974 and the Management of Health and Safety at Work Regulations 1999.

Make sure you are aware of the term 'risk assessment' and what it means to the train drivers role.

The following is a sample response to this type of question.

Sample Response

What do you understand about the term Health and Safety and who is responsible for it?

"Everybody is responsible for Health and Safety at work. Health and Safety is governed by the Health and safety at Work Act 1974 and the Management of Health and Safety at Work Regulations 1999. Train drivers are responsible for the safety of themselves and the safety of each other.

Health and Safety is all about staying safe and promoting good working practices. In the rail industry this means making sure that all protective

clothing is worn when required, following rules and procedures, checking that equipment and machinery is serviceable and carrying out risk assessments when required. It also includes simple things like making sure warning signs are placed out after the train station has been cleaned.

It applies both when driving trains and also when carrying out duties around the train station. Health and Safety should be at the forefront of everybody's minds when at work."

Question 23

How do you keep yourself fit and why do you think fitness is important to the role of a train driver? (Manager's interview)

Fitness is an important element of the train driver's role. Of course, you do not need to be super fit, but a good all-round level of fitness will help you to concentrate for long periods of time. Because the role requires a considerable amount of time sat in a cab you should also keep yourself fit for your own well-being. If you sit in a train cab for hours each day and then go home and sit on your backside for the evening it is not going to do your long-term health any good.

Eating properly is also key to maintaining a healthy lifestyle. Whilst you are unlikely to be asked questions about your diet, you will find that you feel a whole lot better about yourself if you eat properly.

The following is a sample response to this question, based on a person who lives a healthy lifestyle and keeps physically fit.

Sample Response

How do you keep yourself fit and why do you think fitness is important to the role of a train driver?

Yes, I keep myself fit and active and it is an important part of my life. I go swimming 3 times a week and swim 30 lengths every time I go.

I also play football for my local Sunday team, which involves one practice session every fortnight. However, I understand that I may not be able to continue with that sport due to the hours I will be required to work as a train driver. I also ensure that I eat a proper diet, which helps to keep me feeling confident and healthy.

Yes, I think that fitness is vital to the role of a train driver. The role involves working unsociable hours and I understand that it can be stressful at times. You also need to maintain your fitness levels so that you can learn and retain new information and also maintain your concentration levels.

Question 24

How do think you would cope with a fatality? (Manager's interview)

As a train driver there is a possibility that you will come in to contact with a fatality at some point during your career. Personally, I have witnessed many fatalities during my career and it is not a particularly pleasant experience. However, I always maintained a calm mind-set and tried to understand that it was part of my job. You will probably find that the train operating company offer some form of counselling for train drivers who witness fatalities, particularly those which involve suicide. Here's a brief sample response to this question based on how I would personally respond to it.

Sample Response

How do think you would cope with a fatality?

"I understand that there is a possibility that I will witness a fatality during my career as a train driver. I believe I would cope well with it because of my attitude towards work and also my mind-set. I am a confident and resilient person and would accept that this could be part of the job. I also under-stand that I would have to be very mindful of the passengers who were travelling on the train with me and it would be important that I followed my training and procedures during such an incident. I would like to think that I would cope well with the pressure of such a situation."

Question 25

Name the train stations from one route to another. (Manager's interview)

You may get asked a question during the interview which tests your knowl-edge of the routes that you will be required to travel once you become a train driver. I would recommend carrying out some research into the TOC which you are applying for to see which train stations they operate out of and also the different stations in between each route that operate.

FINAL TIPS FOR PREPARING FOR THE INTERVIEWS

- Make sure you turn up to your interview on time! Find out the route to the interview location well in advance and make sure you don't get stuck in traffic or have any problems parking. Prepare for these eventualities well in advance.

- Wear formal clothing for your Interview. Make sure you are clean-shaven and your shoes are clean and polished. Remember that you will be representing the company if you are successful and your appearance is very important.

- Visit the website of the TOC you are applying for and learn information about how they operate and what they are about. This is important so that you can create an image that you are serious about working for them and not just interested in becoming a Train Driver. Find out as much as possible about their geographical locations, where they operate out of, their stations and their depot structure.

- During your preparation for the interview, try to think of some recent examples of situations you have been in that are relevant to the role of a Train Driver.

- When responded to the questions try to concentrate on what you have achieved so far during your life. It is important that you can demonstrate a track record of achievement.

- Make sure you smile during your Interview. Sit up straight in the chair and do not fidget.

CHAPTER 4
USEFUL CONTACTS

Within this section of the guide I have provided you with a list of Train Operating Companies that exist in England, Scotland and Wales. Please note, the list is not exhaustive and you may find other TOCs operating within your area/region. Some of the contact details may also change from time to time.

Arriva Trains Wales
www.arrivatrainswales.co.uk

Arriva Trains Wales
St Mary's House
47 Penarth Road
Cardiff
CF10 5DJ

0845 6061 660
customer.relations@arrivatrainswales.co.uk

C2C

www.c2c-online.co.uk

10th Floor,
207 Old Street,
 London
EC1V 9NR

0845 6014873
c2c.customerrelations@nationalexpress.com

Chiltern Railways

www.chilternrailways.co.uk

Chiltern Railways 2nd floor,
 Western House
Rickfords Hill
Aylesbury
Buckinghamshire
HP20 2RX

08456 005 165

CrossCountry

www.crosscountrytrains.co.uk

CrossCountry
Cannon House
18 The Priory
Queensway
Birmingham
B4 6BS

0870 010 0084
info@crosscountrytrains.co.uk

East Coast

www.eastcoast.co.uk

East Coast House
25 Skeldergate House
York
Y01 6DH

08457 225 225

East Midlands Trains
www.eastmidlandstrains.co.uk

1 Prospect Place
Millennium Way
 Pride Park
Derby
DE24 8HG

08457 125 678
getintouch@eastmidlandstrains.co.uk

Eurostar
www.eurostar.com/

Times House
Bravingtons Walk
Regent Quarter
London
N1 9AW

08701 606 600

First Capital Connect
www.firstcapitalconnect.co.uk

First Great Western
www.firstgreatwestern.co.uk

Head Office
Milford House
1 Milford Street
Swindon
SN1 1HL

08457 000 125

First Hull Trains
www.hulltrains.co.uk

First Hull Trains
FREEPOST
RLYY-XSTG-YXCK
4th Floor
Europa House
184 Ferensway
HULL
HU1 3UT

08456 769 905

First TransPennine Express
www.tpexpress.co.uk

Floor 7
Bridgewater House
60 Whitworth Street
Manchester
M1 6LT

0845 600 1671

Gatwick Express
www.gatwickexpress.com

Go Ahead House
26-28 Addiscombe Road
Croydon
CR9 5GA

0845 850 15 30

Grand Central
www.grandcentralrail.co.uk
River House
17 Museum Street
York
YO1 7DJ

0845 6034852
info@grandcentralrail.com

Heathrow Connect
www.heathrowconnect.com

Heathrow Express
www.heathrowexpress.com

6th Floor
50 Eastbourne Terrace
Paddington
London
W2 6LX

020 8750 6600

Island Line Trains
www.southwesttrains.co.uk

Ryde St Johns Road Station,
Ryde, Isle of Wight
PO33 2BA

01983 812 591

London Midland
www.londonmidland.com

102 New Street
BIRMINGHAM
B2 4JB
0121 634 2040

comments@londonmidland.com

London Overground
www.tfl.gov.uk/overground

London Overground Rail Operations
125 Finchley Road
London
NW3 6HY

0845 601 4867

London Underground
www.tfl.gov.uk/

Merseyrail
www.merseyrail.org

Rail House
Lord Nelson Street
Liverpool
L1 1JF

0151 702 2534

National Express East Anglia
www.nationalexpresseastanglia.com

Floor One
Oliver's Yard
55 City Road
London
EC1Y 1HQ

0845 600 7245
nxea.customerrelations@nationalexpress.com

Northern Rail
www.northernrail.org

Northern Rail Ltd
Northern House
9 Rougier Street
York
YO1 6HZ

0845 00 00 125
customer.relations@northernrail.org

ScotRail
www.scotrail.co.uk

Atrium Court
50 Waterloo Street

Glasgow
G2 6HQ
08700 005151

South West Trains
www.southwesttrains.co.uk/

South West Trains
Friars Bridge Court
41-45 Blackfriars Road
London
SE1 8NZ
08700 00 5151

Southeastern
www.southeasternrailway.co.uk

Customer Services
PO Box 63428
London
SE1P 5FD
0845 000 2222

Stansted Express
www.stanstedexpress.com

Virgin Trains
www.virgintrains.co.uk

Virgin Trains
85 Smallbrook
Queensway
Birmingham
B5 4HA

0845 000 8000

Wrexham & Shropshire
http://www.wrexhamandshropshire.co.uk

The Pump House
Coton Hill

Shrewsbury
SY1 2DP

0845 260 5233
info@wrexhamandshropshire.co.uk

OTHER USEFUL WEBSITES

The following website is the UK's leading Train Driver information website. It is an invaluable source of information and one that I strongly recommend you visit and study:

www.railwayregister.care4free.net/becoming_a_train_driver.htm

Rail Safety & Standards Board
www.rssb.co.uk/

The Department for Transport
www.dft.gov.uk/

Rail Technical Pages
www.railway-technical.com/

Track Access – Route Learning website
www.trackaccess.net

Visit www.how2become.com to find more titles and courses that will help you to pass the Train Driver selection process, including:

• How to pass the Train Driver interview DVD.

• 1 Day Train Driver course.

• Psychometric testing books and CD's.

www.how2become.com

Printed in Great Britain
by Amazon